D0686939

Robert Welch: Design in a Cotswold workshop

OAKEY

Robert Welch

Design in a Cotswold workshop

Introduction by Alan Crawford

Edited by Colin Forbes

Photography by Enzo Ragazzini

Whitney Library of Design
Imprint of Watson-Guptill Publications of New York

Lund Humphries London

First edition 1973
Published by
Lund Humphries Publishers Limited
12 Bedford Square London WC1

SBN 85331 365 2

Designed by Pentagram Design Partnership
Made and printed in Great Britain by
Lund Humphries, Bradford and London

Design in a Cotswold workshop

We went over the silk mill again, measured it up, peered through the green and bottle glass panes, studied the girth of the plum tree growing round the stonework, tipped the old mad woman with the ringlets, climbed up and down the empty 17th century houses, asked endless questions and finally in the words of the exceedingly stolid Bill Thornton, the foreman of the smithy, professed ourselves as 'very agreeably surprised'. There now, the country has charms after all, and it seems as if the Great Move were at last coming off.

The quotation is from the journal for 14 November 1901 of Charles Robert Ashbee, architect, designer, and social reformer. He had been with the foremen of his workshops to inspect the old silk mill in Sheep Street, Chipping Campden, in the Cotswolds. The 'Great Move' was the proposed transfer of the Guild of Handicraft, as his workshops were known, from the East End of London to this quiet town.

The revival and vitality of traditional craftsmanship in an industrial society is a remarkable thing. In the past hundred years or so, craftsmen and designers of this new breed have seen their place in the world of design and industry in many different ways, and one of the most interesting variations on this theme is the work of C. R. Ashbee. He was an architect by profession but a large part, perhaps most, of his time was spent as director and designer of the Guild of Handicraft. He started the Guild in 1888 with four workmen producing furniture, beaten metalwork, decorative painting, and modelled work. During the following decade, he added workshops for silver, jewellery, enamels, wrought iron, wood-carving, and printed books. As the Guild expanded it needed larger workshops, and in 1891 it moved from its original premises in the Commercial Road to a large eighteenth-century building in the Mile End Road, called Essex House.

As its name suggests, the Guild was a rather special institution. For instance, it had a co-operative structure: not all the workmen were members of the Guild, but all were free, after a certain period of service, to join it, and when they did so they took out shares and were able to attend the committee meetings of the Guild which decided questions of policy. Another distinctive feature was its size – it employed as many as seventy workmen in the late 1890s – and the range of its workshops – it was practising ten different crafts by 1898. This size and range reflected Ashbee's beliefs, his particular version of Arts and Crafts ideals.

The revival of crafts as a conscious alternative to industrial production really stems, in this country, from the Arts and Crafts Movement of the late nineteenth century, of which Ashbee's work was a part. This movement was dedicated to the revival of craft traditions as a direct challenge to the existing system of industrial production. Within that broad aim, there were different motives and different emphases. From one point of view, the purpose of the movement was to improve standards of workmanship, to teach an understanding of the nature of materials, construction and design. From another, the work produced mattered less than the man who produced it; the value of craft production was that it was a better kind of work, a more creative, humane and pleasant experience.

These were twin ideals in the movement, and most of its members held to them both. But if there was one thing which marked out Ashbee, it was the stress he laid on the second point of view. For him, more than for most others, the Arts and Crafts Movement was a social as well as a design experiment. He felt that what he was creating in the first place was a better way of life and work, and beautiful objects only after that. It was partly for this reason that he built up the Guild in the 1890s. If craftsmanship was a better kind of life, then as many people as possible had to be craftsmen.

Portrait of Ashbee modelled in relief in plaster. Possibly designed and made by Ashbee 1891

He had to spread the benefits of the experience of working in the Guild as widely as possible. His attitude was anti-industrial, but not escapist; rather, it was outward-looking and aggressive.

To expand the Guild, and still to keep the atmosphere of a small workshop rather than a factory meant adding different craft shops as separate units to the Guild's range; and this he did throughout the 1890s. He acted as designer for all these crafts, though he had no manual skill himself except in drawing and modelling; in the Guild, design was a collaborative process. Ashbee's design on paper would be fairly explicit, but it could be modified at a later stage in discussion with the craftsman concerned. Besides, some workmen were encouraged to carry out their own designs; and there were some Ashbee designs that were so much in

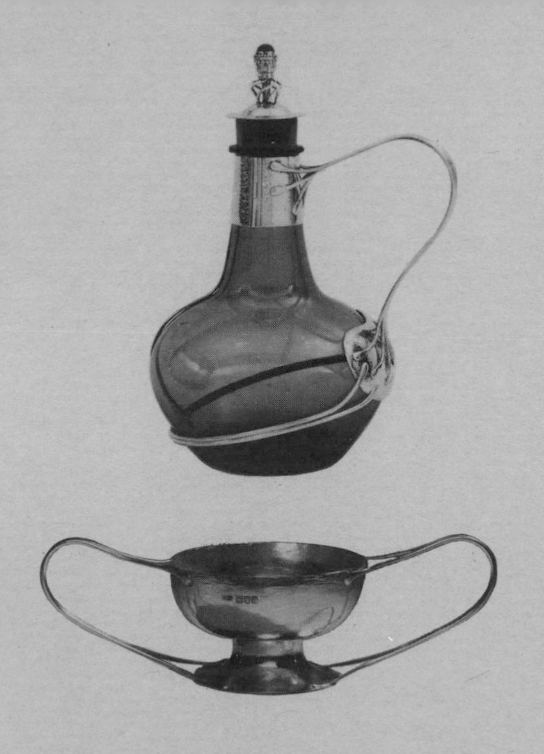

Silver-mounted decanter and Silver butter dish designed by Ashbee and made by the Guild of Handicraft 1901 (*Crown copyright, Victoria and Albert Museum, London*)

demand that they became standard Guild products and could be produced without his intervention. The two-handled jam and butter dishes and the silver-mounted decanter are cases in point. Almost all of Ashbee's designs were for individual pieces, costly in labour and often in materials; he did in fact design for industry, but the only surviving evidence of these designs is an article about some fireplaces which he designed for the Falkirk Iron Company. Much of the Guild's work was commissioned, and the retailing of the rest was a fairly unsophisticated affair. They relied on showrooms at the workshops in the East End until 1898, when a Guild shop was opened at 16a Brook Street, just off Bond Street. When the Guild moved to Campden in 1902 a gallery was opened at 67a Bond Street. These steps were taken on the advice of friends, and the truth is that Ashbee was ignorant and perfunctory when it came to selling the goods his workmen had made. He despised 'shopkeeping' as far inferior to the creative work of the craftsman.

Much as the Cockney atmosphere of the East End appealed to Ashbee, he felt that it did not provide congenial surroundings for the sort of workshop which he was trying to build up, one in which work would be pleasurable, and leisure would be instructive. Increasingly, he became convinced of the necessity of moving the workshops out into the country.

The lease on Essex House ran out in 1902, and Ashbee inspected sites in Kent, at Letchworth, and finally at Campden, attracted by the possibilities of the old silk mill, and by the architectural distinction of the town, a distinction not altogether marred by the dilapidated condition of some of the houses, for there seemed fair scope here for the talents of an architect who specialized, as Ashbee did, in restoration work. Inspections, such as that recorded in his journal, were made, and on Christmas Day 1901 he was able to record that the Guildsmen in committee had decided, as he put it, to 'leave Babylon and go home to the land'.

In his enthusiasm for the place Ashbee thought that he had discovered Campden. In fact, several of his friends knew it already, and well before 1902 the Cotswolds had begun to draw artists and architects from London. They were fascinated by the quality of its vernacular building, a quality so closely fitting the ideals of the Vernacular Revival in architecture that it almost seems as if the emphasis set by that revival on simple domestic building and on the sensitive use of local materials and techniques was formed with the Cotswolds in mind. The most famous of the Cotswold migrants were Ernest Gimson and Ernest and Sidney Barnsley, archetypal furniture designers of the Arts and Crafts Movement, who settled near Cirencester in 1893. In 1902 they opened their Daneway Workshops, manned in part by local craftsmen, and in 1903 built houses for themselves in Sapperton. The architecture of the village was gradually transformed by their presence.

Chipping Campden had not experienced any such transformation, and in the spring of 1902 was bracing itself for something like an invasion. Structural repairs were carried out at the mill, and an electric power plant, the first in Campden, was installed. It was possible to fit all the workshops comfortably into the building, except for the smithy, which had to go in an outhouse because of the noise. The arrangement of the workshops in the mill was, from left to right: Ground floor: showrooms, Essex House Press, drawing office, and general offices. First floor: jewellery, silversmithing and enamelling. Second floor: cabinet making, wood carving and french polishing.

In the 'Great Move' as many as 150 men, women and children settled in Campden, which at that time had a population of about 1500. It is easy to imagine the difficulty with which the people of Campden adjusted to this sudden addition to their number. The way of life and expectations of the East Enders were very different from theirs. For some time there were two prices in the shops, a Guild price and a Campden price, so sharply was the difference felt between the natives and the outsiders. The distinction was not made any less by Ashbee's well-intentioned efforts to show everyone in Campden, ex-Londoner or not, how to take advantage of their situation. He brought to the Campden experiment his habitual energy and idealism, and set about transforming Campden according to his ideal of a revitalized country life.

He founded the Campden School of Arts and Crafts, where he taught Campden's young men to be healthy and her young women to wash and cook. The Guildsmen built a swimming pool near Westington Mill and held water sports there, beating all comers; and at Christmas each year, they performed plays, mostly Elizabethan dramas acted in the old way with little scenery. One participant recalls that they were 'appreciated but moderately, I think, by the audience, but vastly enjoyed by the performers'.

These mock-rustic antics must have seemed strange to the inhabitants, and it was not by reviving a supposed vernacular culture that Ashbee changed the town, but rather by the links he forged between Campden and London. It was not the press-ups which counted at the Campden School of Arts and Crafts, but the loan exhibits of decorative art from the Victoria and Albert Museum in the school's gallery, as Sir Gordon Russell recalls in his autobiography *Designer's Trade*; it was the lectures at the school by artists like Walter Crane and the presence in Campden at weekends of such guests as John Masefield, Sidney and Beatrice Webb, and Laurence Housman, which really brought a new element into Campden life. The visitors' book still in use today in the mill is signed in September 1910 by Frank Lloyd Wright.

A year or two after the Guild arrived, other artists began to settle in Campden, following Ashbee's lead. F. L. Griggs, who was an architectural draughtsman when he came to Campden, making his living with lyrical, sharply detailed perspectives of other architects' work, drove into the town in 1904 on one of the first motor bicycles ever seen in Campden, to do the illustrations for the 'Oxfordshire and the Cotswolds' volume of the *Highways and Byways* series of topographical guides. He stayed for the rest of his life, living until after the First World War in Dover's House in the High Street, and then in Dover's Court, the house he built for himself between the High Street and Back Ends, one of the last and one of the most painstaking small country houses of the Arts and Crafts Movement. At about the same time Paul Woodroffe, book illustrator and stained glass artist, set up his studio in what is now Woodroffe House, Westington, restored by Ashbee. And in 1907 Ananda Coomaraswamy, the authority on India art and religion, settled in Broad Campden where Ashbee restored and added to the Norman Chapel for him. (Coomaraswamy's wife, Ethel, married a second time, and, as the weaver Ethel Mairet, is a well-remembered figure from British design in the 1930s.) None of these artists became members of the Guild of Handicraft, but all of them worked closely with it and would perhaps not have come to Campden had it not been for the Guild. In this way Ashbee and the Guild began to change

'he Guild of Handicraft Workshops at Chipping Campden. From an old photograph

'hysical fitness classes at the Campden School of Arts and Crafts. From Craftsmanship in Competitive Industry, *C. R. Ashbee 1908*

Campden from a quiet town preoccupied with farming to one, hopefully at least still quiet, but now concerned with farming, craftsmanship and art.

They did this despite the fact that the Guild's effective life in Campden as a limited company lasted only five or six years. In about 1905 it began to run into financial difficulties. The causes are not all clear, but it is obvious that the bold, idealistic step of moving the workshops out to Campden was partly responsible. Ashbee employed a large number of skilled workmen, and in London some of them could easily find temporary work outside the Guild when work was slack, and so could be laid off without difficulty. But in Campden, when work was slack, as it was from 1904 onwards, such workmen had to be kept on, making goods for stock at a time when prospects for sales were already poor. The distance of Campden from London was another disadvantage: it added to transport costs, and loosened the link between the Guild and its clients. Drastic measures were taken in 1906 to try and improve the situation; at the end of 1907 the Guild of Handicraft Limited went into liquidation.

As a result, many of the Guildsmen had to leave Campden and find work elsewhere. But Ashbee himself remained there, working as an architect and designer, until the First World War, and did not leave Campden for good until 1919. Also, some Guildsmen went on working in the mill, now under their own names. Jim Pyment, who had been foreman of Ashbee's cabinet-making shop, bought the mill, and started a building firm, J. W.Pyment and Sons, which has been responsible for much of the careful and traditional building work in the area. The firm is now run from the ground floor of the mill by his son, Harold Pyment. The top floor of the building was taken over by Miller and Hart, architectural sculptors and carvers; that is, Alec Miller, a carver who joined the Guild from Glasgow in 1902, and Will Hart. This shop lasted until 1939, when Alec Miller left for the United States to pursue an active career as a sculptor. Charlie Downer and Bill Thornton, the principal Guild blacksmiths, took over the smithy under their own names, and were active until the 1950s.

The first floor continued to be occupied by silversmiths, and one of these was George Hart. He joined the Guild in 1901, after his metalwork had been noticed by Ashbee at an exhibition of Arts and Crafts work in Hitchin; and his brother Will, the carver, joined at the same time. Before this, he had fought in the Boer War, and had had some experience of farming, which stood him in good stead after 1907.

After the liquidation of the Guild, Ashbee devised a scheme to give his workmen greater economic security. He persuaded Joseph Fels, the millionaire American inventor of naphtha soap, and an energetic propagandist for land nationalization, to finance the acquisition by the Guild of seventy acres in Broad Campden. Several of the Guildsmen started smallholdings on this land, but as it turned out only George Hart had the experience to take advantage of the scheme. Ever since, he and his family have been farmers and craftsmen, much as Ashbee originally intended.

In the first years after liquidation, George Hart worked with his step-brother Reynell, under the name of Hart and Huyshe; since then the continuous tradition of his workshop has been maintained; and though now, at the age of 91, George Hart no longer works at the bench, the shop is run by his son and grandsons. They work with the same simple hand tools used in the Guild, and their work is traditional in the best sense: much of George Hart's work has been done for churches in the Cotswolds and abroad.

After the First World War, life in Campden went on very much as before. The range of work produced in the Guild – as the workshops were still known – and by other Campden artists can be seen in the catalogues of the Arts and Crafts Exhibitions held in Campden from 1924 onwards. In the foreword to the 1924 catalogue, Wentworth Huyshe, heraldic artist, and stepfather of the Hart brothers, wrote: 'The Campden artists and craftsmen, wishing to perpetuate the reputation of the town as an art centre, and to continue as far as possible the work of the Guild founded by Mr Ashbee, have conceived the idea of holding an exhibition of works produced by resident artists and craftsmen of whom no fewer than four have works in this year's exhibition of the Royal Academy.'

Among the exhibitors that year were Arthur and Georgina Cave Gaskin, the Birmingham jewellers, who had a cottage in the High Street; and it is known that other artists of the Birmingham group, the stained glass artist Henry Payne, and the painter and illustrator Charles March Gere, were frequent visitors to Campden. The catalogue of the exhibition was printed at the Shakespeare Head Press, Stratford-on-Avon, and at this date the links between Campden artists and Bernard Newdigate and Basil Blackwell must have been strong. Campden, in any case, had a quality press of its own from 1928, when H. P. R. Finberg set up the Alcuin Press in the house Ashbee had used for his architectural office.

During these years, the importance of Campden as an art centre was eclipsed by its neighbour, Broadway. There Katherine Adams had set up her bindery before the First World War. After it Gordon Russell developed, from the repair shop attached to the Lygon Arms and his father's antiques business, a workshop producing furniture to his own designs and employing local craftsmen. The early furniture harks back to the characteristic Arts and Crafts style of Gimson and the Barnsleys; but this was only a point of departure, and the distinguishing mark of Gordon Russell and his firm has been a power for development. From round about 1930, Gordon Russell Limited was producing work comparable with that of Modern Movement designers on the Continent; and from the same date the firm was increasingly involved in design for quantity production, starting with the production of cases for Murphy Radio in 1930. The firm quickly grew to a size never contemplated by Ashbee in Campden – 120 workmen in 1928 – but it maintained the principle of craftsmanship, making furniture designs in very small batches. Gordon Russell's active part in design work came to an end in the mid 1930s, and in 1947 he became director of the newly-founded Council of Industrial Design. The firm had been a member of the earlier Design and Industries Association since the mid 1920s, and in this latest and widest phase of his career, he was able to carry on its campaign to promote the quality of craftsmanship in industry. Still Chairman of the Lygon Arms and of Gordon Russell Limited, Sir Gordon lives on Dover's Hill, overlooking Campden. His place in this story is to have brought to the Cotswolds design as we know it today.

In the summer of 1955, Sir Gordon had a letter from Robert Gooden, professor of silversmithing at the Royal College of Art, on behalf of a young silversmith who had just finished at the College and was looking for a studio-workshop within striking distance of Birmingham. He thought of his friend George Hart, and of the Guild workshops in Campden. As it happened, the top floor had been empty, used only as lumber room, since Alec Miller had gone to America. Soon a three-years' lease had been arranged between the owner, Harold Pyment, and the silversmith, Robert Welch.

During the summer the workshop was fitted out, and a small space was partitioned off from the workshop, to act as a studio-cum-office-cum-bedroom. At this time, the whole of Robert Welch's workshop filled only the northern end of the building, but gradually it has expanded to occupy the whole of the top floor, and this expansion has sometimes uncovered traces of the Guild. Plaster casts for statues by Alec Miller turned up in 1968, and later a medallion portrait of Ashbee, possibly a self-portrait. These accidental finds are typical of the way in which Robert Welch, who had never heard of Ashbee – Jackson's *Illustrated History of English Plate* was the set book for history at Birmingham in the early 1950s, not Pevsner's *Pioneers of Modern Design* – has stumbled upon the earlier craft traditions of Campden. For several years he heard talk of the Guild and of Ashbee, but always in the tantalizing form which oral tradition takes: the past and its figures idealized, but much that was important overlooked, for the very reason that it was important. In the end he decided to find out for himself, and spent some time in the Library of the Victoria and Albert Museum, reading Ashbee's *Memoirs*. He now has a growing collection of silverwork, printed books, furniture and ephemera relating to the Guild, from which some of the illustrations for this book are taken.

Unlike Ashbee, Robert Welch was not a complete stranger to the Cotswolds when he came to Campden. He was born not far away, in Hereford, in 1929, and was brought up in and around Malvern. In 1946 he went to Malvern Art School to study life drawing and painting, and it was at this time that he first became interested in metalwork. His course was interrupted by National Service, but in 1949 he returned to Malvern to take the intermediate examination for the National Diploma in Design, and then applied to complete his diploma at the Birmingham School of Art, where metalwork under Ralph Baxendale and Cyril Shiner was taught to a high standard. Despite this standard, there were only two other silversmiths in Robert Welch's year, probably because of the very poor prospects in silversmithing at that time, when purchase tax was at a rate of about 100%. Robert Welch has always been an individualist, and in these years he needed to be. When, at the end of his second year at Birmingham, he was accepted for further study at the Royal College of Art, he found that he was the only silversmith in his year.

Nevertheless, his years at the Royal College, from 1952 to 1955, brought him into contact with other silversmiths, especially Gerald Benney and David Mellor. At the Royal College they were at the heart of a revival of silversmithing, as can be judged from the fact that they were the first to set up independent craft workshops as silversmiths when they left the College. Before 1955, anyone leaving a college of art as a trained silversmith would automatically have gone into teaching, in order to carry on silversmithing on the side. One has really to go back to the time of the Arts and Crafts Movement to find craft workshops springing up in this way, as centres of advanced design.

Perhaps the most important thing that happened to Robert Welch at the Royal College was stainless steel. In the summer vacation of 1953, with a £50 travelling scholarship which he had won from Birmingham the year before, he went to Sweden for a design course organized by the Swedish Council of Industrial Design; in 1954 he went to Norway and worked with the silversmith Theodor Olsen of Bergen (where he made a silver flower vase which was bought by the Bergen Museum of Applied Art). Scandinavia brought stainless steel to his attention, and in his third and last year at the Royal College he specialized in stainless steel. His diploma thesis was entitled 'The Design and Production of Stainless Steel Tableware', and drew on those Scandinavian examples: the illustrations were all taken from the work of designers like Folke Arstrom, Erik Herlow and Sigurd Persson, except for three designs by Harold Stabler from the mid-1930s.

Robert Welch's main project in the final year was an entrée dish with a cover and removable triple liner. As a prototype, it had to be made in gilding metal and satin chrome plated, but it was an experiment in design for stainless steel, and, as such, it was a success. At that date, the only manufacturers of stainless steel tableware were J. & J. Wiggin, a small family firm at Bloxwich, just north of Birmingham, whose products were marketed under the name 'Old Hall'. They were impressed with Robert Welch's entrée dish, bought the design, and offered him a post as design consultant, with the requirement that he should spend one day a week at their factory in Bloxwich. So it was that, in the summer of 1955, Robert Welch was on the lookout for a studio-workshop in the Midlands.

There is a curious contrast between Robert Welch's reason for settling in the mill at Campden, and that which inspired Ashbee fifty years earlier. For Ashbee the 'Great Move' was an escape from the city, from London, into a district whose whole attraction was that it was quiet, out of touch, an industrial backwater. Robert Welch may have chosen Campden with some of this feeling; but for him the chief attraction of Campden was that it was near to a city, and to an industrial centre. This is symptomatic of the difference between Ashbee's approach to craftsmanship and Robert Welch's: for this latest occupant of the mill, craftsmanship is integrated with, runs fruitfully alongside, design for industry. For Ashbee Campden meant the atmosphere of a pre-industrial way of life, and, sadly, financial failure owing to poor communications, getting out of touch. For Robert Welch in the 1950s Campden meant, and still means, a congenial base for work that fits in in various ways with industrial production, and happily, thanks to good communications, keeping in touch.

Work for Old Hall, which originally brought Robert Welch to the area, has remained his biggest single commitment in industry, and to understand this, something of the background, and especially Old Hall's unique concern with stainless steel, needs to be explained. 'Olde Hall' was the brand name of J. & J. Wiggin Limited, originally the local name of the old Salvation Army hall in Bloxwich which they used as a workshop at the turn of the century. During the first thirty years of their existence, from the 1890s onwards, they produced a variety of metalware, belt buckles, stirrups, bathroom fittings, and so on. In the 1920s they turned to stainless steel as a suitable material for bathroom fittings; and in 1928 they were the only firm to respond to the programme sponsored by Firth Brown of Sheffield to popularize stainless steel in domestic use. In 1934 they exhibited the world's first stainless steel tea set in 'Staybrite City', Firth Brown's exhibit at the *Daily Mail* Ideal Home Exhibition. This and succeeding pre-war tea and coffee sets were designed by the Wiggin family.

As part of their programme to promote stainless steel, Firth Brown had appointed a designer, Harold Stabler, whose work in this field makes yet another link with the earlier craft traditions. Stabler's work and career reflect accurately the development of English design in the early twentieth century. Born in 1872, he was trained as a metalworker at the Keswick School of Industrial Art, which had been founded by the 'Apostle of the English Lakes', Canon Rawnsley and, perhaps inspired by the lonely presence of Ruskin by Coniston Water, became an outlying centre of the Arts and Crafts Movement. In his early twenties, Stabler moved to London, and settled in Upper Mall, Hammersmith, a domestic haven of the Movement ever since William Morris had moved there in 1878.

His reputation was established during the first decade of the century, especially by his cloisonné enamels, and other metalwork shown at Arts and Crafts Exhibitions. Just before the First World War he seemed to take a new direction. Sir Nikolaus Pevsner says that he was the 'most active' of those who saw in the failure of the 1912 Arts and Crafts Exhibiton a sign that the vitality of the movement had gone, and that a renewed effort would have to be made, especially in the direction of bringing together design, craft and industry, something that the Arts and Crafts Movement with its hostility to industry, could not do. This sense of discontent gave birth to the Design and Industries Association, and Stabler was one of its founders. The Association saw its work in part as the bringing together of designers and manufacturers; and it probably explains why heraldic tiles by Stabler can be seen in some of London's Underground stations, and why, in the 1930s, Stabler should have been approached by the principal manufacturers of stainless steel and asked to design as wide a range of products in the material as he could.

As part of their agreement with Firth Brown, Old Hall developed a range of tableware designed by Stabler. He used an elaborate technique of etching on stainless steel, to counter the effect of scratching, and his work as a designer is generally remembered at Old Hall as almost pedantically precise. The importance of his work lay less in the designs – they could only be got on to the market in 1939 and were not a success – but in the fact that a designer should have been used for a material which at that date lacked any prestige in domestic use.

During the Second World War, Wiggins were attached to the Admiralty for war work, and all production of tableware was dropped. This left them in an unfavourable position after the war, because Scandinavia, where there had been no commitment to armaments, had forged ahead in the design and production of stainless steel for domestic use. Whereas Scandinavia had been unknown in this field before the war, they led the world in design and production after it, as Robert Welch had recognized while at the Royal

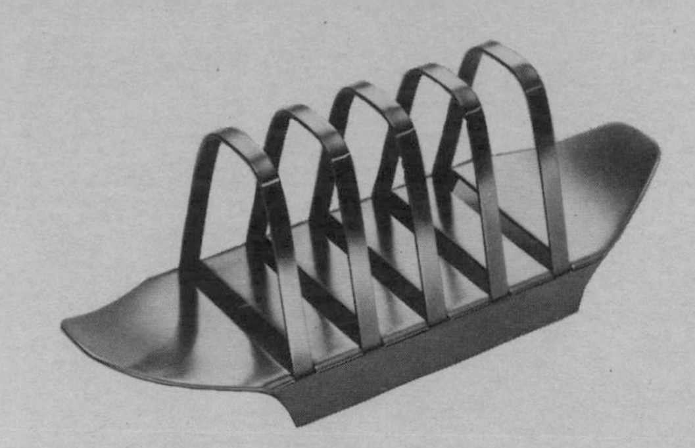

Stainless steel toast-rack 1956

College. During these years Old Hall reissued Stabler's design in a form modified by Leslie Wiggin, who has with versatility been both managing director and designer at Old Hall until recently. But, in the new situation of the mid-1950s, they realized that the firm could make use, once more, of an independent designer who would develop a competitive style, a distinctive look for British stainless steel, so they appointed Robert Welch as consultant.

An Old Hall/Robert Welch design for a toast rack was produced within a year of his appointment. It was effective in attracting attention, and began the process of changing the image of British design in stainless steel. In 1957 the process was taken a stage further, when the Council of Industrial Design suggested to Old Hall, and to Walker and Hall of Sheffield, that they should produce a range of cutlery which would compete with the Scandinavians. David Mellor was the design consultant for Walker and Hall, and so he and Robert Welch produced the 'Campden' range, which was manufactured by Walker and Hall and marketed by Old Hall. The design of 'Campden' is described by Leslie Wiggin as 'middle of the road'. The challenge to the Scandinavians, and especially to the popular 'Facette' design by Folke Arstrom, manufactured by A. B. Gense of Sweden, had to be married with practicality and a kind of ordinariness.

At the time that 'Campden' was being developed, Old Hall received vindication of their choice of Robert Welch as design consultant. In 1958 the Council of Industrial Design put him in touch with the Orient Line, who asked him to take charge of the design of tableware for their new liner *Oriana*. At this stage the shipping line had no definite idea of what they wanted; so Robert Welch was able to suggest that they try the experiment of stainless steel instead of electroplate, and that the firm for which he worked as a design consultant would be interested in manufacturing his designs. The proposal was accepted, and a very large order, involving the design of many different items, was placed with Old Hall, who were also able to issue the design later on the retail market.

Oriana Jugs 1958

The most important work which Robert Welch has done in stainless steel since that time is illustrated and explained in the body of the book. His most recent work, which has not yet appeared, is a new range of cutlery and a tea service. This work has been done against the background of important changes at Old Hall. In 1970 the company was taken over by the Prestige Group, and in 1971 Leslie Wiggin retired as managing director. At the same time, the increased competition in this field has meant a shift in design emphasis. While Robert Welch's relationship with the firm has not changed, the accent in this most recent work has been changed somewhat to meet specific marketing requirements.

Football Manager of the Year trophy 1965

In the early days of Robert Welch's career, work for Old Hall provided a mainstay. But it was only one side of his work. His principal intention, after all, in setting up his workshop in Campden was to practise as an independent silversmith, and during that first autumn of 1955 he was working on two commissions, both of which came to him by way of the Goldsmiths' Company. The role of the Company, and especially of its 'Art Director' Graham Hughes, has been central in the recent revival of design in silver. Graham Hughes has acted as a bureau of information and publicity for designer-silversmiths, and has been able to place many commissions with young designers who might otherwise have gone unnoticed. In fact, Robert Welch's major work in silver during this period was probably the seven-light candelabrum commissioned by the Goldsmiths' Company for the Smithsonian Institution's travelling exhibition of British crafts.

Craft silversmiths have depended on this kind of work, making objects for presentation and display, which have traditional associations, chalices, candlesticks, bowls of various kinds. Modern trophies perhaps belong to this category, but they pose rather special problems. It is not that there are too many traditional associations, but too few. The silversmith has to build into his design some suggestion of the achievement for which the trophy is awarded. Robert Welch did that for the Football Manager of the Year.

Another important job at that time was the 1960 ink-stand commissioned by George Tarratt Limited of Leicester. Robert Welch's commentary on this piece points to the constant cross-fertilization between craft and industry in his work. The inkwells were not produced by the traditional method, but by industrial machining, producing a very high quality of precision and finish. This is what David Pye has called 'the workmanship of certainty'. Robert Welch's feeling for this quality, and the way in which he brings it into his craft work, the traditional home of the 'workman-ship of risk' can be seen also in the 1972 oval teapot with its flush hinge.

This craftsmanship is experimental: Robert Welch will not limit himself to the range of effects which can be produced by traditional craft techniques, and if he wants a particular effect he is prepared to search and experiment until he can get it. This has often meant borrowing or adapting industrial techniques, some-thing that a traditional craftsman would not find it so easy to do. The 1969 Goldsmiths' Hall candelabra are a good example. Here the spherical forms could only be produced by an old swaging machine, originally designed to make watch parts.

Another, less extraordinary, borrowing can be seen in the range of cast-iron decorative pieces. Robert Welch came upon the distinctive disk motif of these designs while experimenting with cast iron at the Beech Hill Foundry near Wolverhampton. He made a small set of candleholders and was so pleased with them that he made some more and sold them to Heal's and Liberty's. In 1964 the press-type nutcracker was added to the candleholders, and other items in the following year. For a short time Robert Welch marketed them himself with the help of his secretary, but their popularity soon made this impossible and in 1963, on the advice of Heal's, he put them in the hands of Wigmore Distributors. In 1968 they were transferred to Old Hall.

These cast-iron pieces have played an important part in Robert Welch's career. It is not only that they were the first pieces to carry his name – they have been marketed as 'Robert Welch Designs' – it is that the contacts made in the course of marketing them have moved the range of his work as a designer onto a different level.

For Robert Welch is not just a silversmith and a designer of stainless steel; he is also a product designer with an international reputation. His first contact with product design came in 1957 when he was asked by A. E. Halliwell of the LCC Central School to act as a visiting lecturer on stainless steel in the School of Industrial Design. Robert Welch brought his practical knowledge of workshop techniques to complement the largely theoretical teaching at the Central School. Conversely, he took away a growing knowledge of drawing office procedure which it had not been the job of the silversmithing workshop at the Royal College to give.

This exposure to some of the techniques of industrial design led to Robert Welch's decision in 1959 to extend his repertoire as a designer in this direction, beyond stainless steel; and from 1961 to 1966 there was a policy of expansion in product design. Important early work included designs for Westclox, which lasted until 1966, sanitary fittings for British Rail (1961–3), redesigning a knife-sharpener (1964) and a complete range of canteen crockery for the then Ministry of Public Buildings and Works (1964).

The marketing of the cast iron greatly expanded these design opportunities. In the United States it led to the H. E. Lauffer cast-iron ware commission, and the link with Wigmore Distributors led to the lamps for Lumitron. The kettle and other kitchenware for Carl Prinz A.G. was commissioned after their managing director had seen the cast-iron ware and Robert Welch's silver on a visit to England in 1965. Perhaps the most gratifying result of the popularity of these designs was the recognition given in Scandinavia. The cast-iron range was particularly well received in Denmark, and in 1966 Robert Welch went on a sales tour of Scandinavia which led in turn to a close association with Skjalm Peterson. In 1967 Peterson held a one-man show of Robert Welch's work in his shop in Copenhagen, and *Mobilia*, the Danish design magazine, published an article on his work in its August number. This in turn led to a commission for a kettle which is being tooled up at the moment by the Wejra Company in Denmark.

By a coincidence, at just the same time, Robert Welch was given a one-man show in London, when a large part of the ground floor of Heal's was given over to a survey of his work. And it was Heal's who were responsible for the major craft project of the mid-1960s. In 1963 they decided to try to establish the habit at the upper end of the market of buying craftsman-made silver of modern design for wedding presents and so on. They asked Robert Welch to design and make a range of silver tableware for them. The whole

Cabinet making workshop of the Guild of Handicraft

The workshop as it is today
(Photograph by Richard Einzig)

range was displayed in a special showcase, where customers could commission a set, or parts of it, from the Campden workshops. The idea was a new one, and response was steady but gradual. In any case, the idea was probably more important to silversmiths than to the public. Here, in principle, was a way in which craft silversmiths could make their work known outside their own shops. Until such a move was made, silversmiths remained dependent on commissions from institutions such as the Goldsmiths' Company or churches, which, because of their traditional requirements, know where to find them. What Heal's did was to give silversmiths a retail platform like any other manufacturer.

The lessons learnt from this London venture, and also from the experience of organizing the marketing of his cast-iron ware, must have weighed in Robert Welch's mind when he decided in 1969 to open a shop to sell his own designs. He bought a house, originally a seventeenth-century beer-house, on the corner of the High Street and Sheep Street in Chipping Campden, only a few yards from his workshops and opposite Elm Tree House, which Ashbee had used as his architectural office. It was converted to its present use during that year, and opened in December; at the time of writing, an extra space has just been added for temporary exhibitions of the work of other artists and designers. The shop sells Robert Welch designs such as Lumitron lamps and Lauffer cast iron; but its main purpose is to be a shop-window, not for the products which can be bought elsewhere, but for his silver. It makes permanent the retail platform originally devised by Heal's.

It is true that Chipping Campden High Street is not the Tottenham Court Road. Campden is not on a main road, and it has been transformed by tourism, at least superficially, rather less than Broadway or Moreton-in-the-Marsh. But the shop has been a success. The tourists and visitors clearly come in sufficient numbers or with the right intentions, and the difference is felt in the flow of orders into the workshop. So times change. When Ashbee was in Campden, parties from art colleges came occasionally to visit the workshops. But a shop such as Robert Welch has set up would not have made sense. Even with the two shops off Bond Street, to catch the customers, the Guild of Handicraft was still out on a limb.

The shop has brought not only customers, but also a closer contact with the public. Before 1969 Robert Welch's silver was almost always carried out at a distance from the customer, and the customer was usually an impressive and impersonal body, the committee or representative of a firm, college or church. Since opening the shop he has found that ordinary members of the public have come in to have work done and have been ready to discuss the work in a relaxed atmosphere. The upshot has been a series of new designs for simple domestic silver which Robert Welch would not have come to but for this fruitful contact. The oval teapot is an example of plate made in this way, after discussion with a customer.

The opening of the shop, with the possibility of making silver for sale, and of encouraging commissions from the public, has meant in the past few years a greater concentration on the workshop and on craft work. This concentration does not imply any cutting back in design. The opportunities which existed in the early sixties, when it was possible for silversmiths such as Robert Welch or David Mellor to branch out in many directions in domestic design, such was the demand for it, have given way to a more specialized and competitive situation. But Robert Welch's own commitment to design continues: Lauffer and Lumitron designs belong to this phase of his career.

All it implies is another way in which craft and industry relate in Robert Welch's work. Here it is not simply a question of these two interests running happily alongside, or of borrowing from one to solve problems in the other. There is a sense in which craft and industry are opposite for Robert Welch. The craft workshop is so completely different. He has always kept his workshop staff small – there have never been more than three craftsmen – and they form a close-knit working group. John Limbrey, who is responsible for the high quality of much of the craft work, has been working with Robert Welch now for fifteen years. Robert Welch's own skill as a craftsman enables him to work closely with the rest of his workshop; and he is free to develop personal designs less trammelled by the requirements of cost and standardization. It is only in this situation that a personal *tour-de-force* such as the large wire-decorated dish made in 1972–3, designed by Robert Welch and made by John Limbrey, is possible.

Craftsmanship and industry, silver and product design, the workshop and the international market: these are twin poles in Robert Welch's work. They provide a kind of dialectic of progress. At one time, perhaps for several years, he can concentrate on product design, allowing silver to run along with the impetus of an earlier effort, taking work as it comes along, but not going out to create it. Then he will find himself spent in the direction of design, and will feel the need to get back to the simple situation of the workshop where he can design more for himself and control production from start to finish. Then the industrial clients will not be turned away; but there will be no effort, as there was in the mid-sixties, to go out and find them. It is at this point that Robert Welch finds himself now.

Alan Crawford
June 1973

Robert Welch 1958–1973

Candelabrum in sterling silver, height 395mm, commissioned in 1958 by the Worshipful Company of Goldsmiths for a travelling exhibition sponsored by the Smithsonian Institution in the USA. The candelabrum is composed of undulating forms, designed to create a shimmering effect when the candles are lit, a feature that was to be used again fourteen years later for the Goldsmiths' Hall candelabrum.

The design was created by the turning of the wooden study models of the seven stems in a lathe in a random configuration of undulating shapes. They were assembled by using horizontal members also made from wood. The study model was then used as the basis for the development of the silver candelabrum. A further set of stems were turned in wood together with the candle-sockets and silver castings were made from these models. The castings were filed and polished, and the candelabrum was soldered together with cross members cut from thick silver sheet.

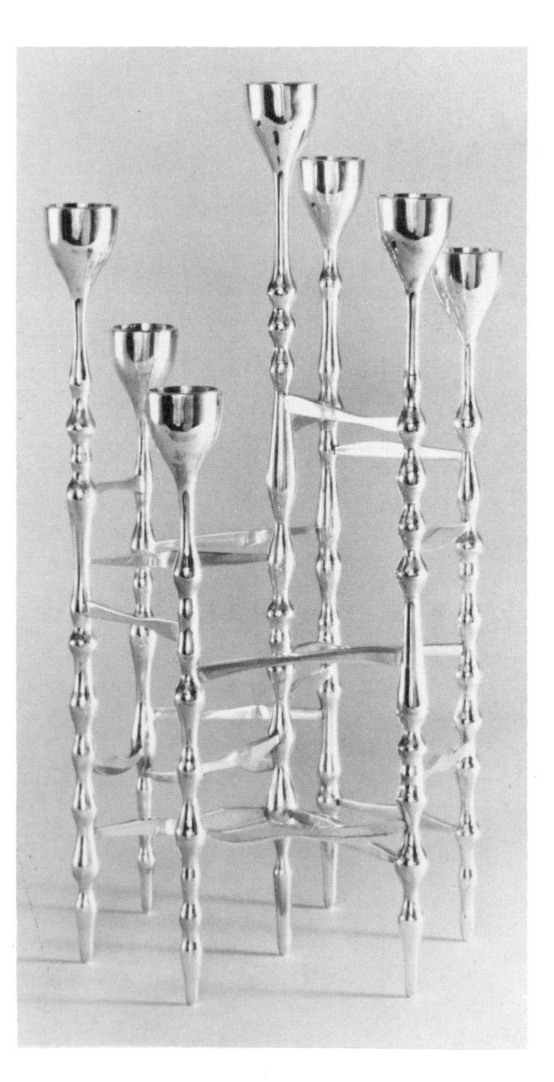

Two glass inkwells, square in plan, are set either side of a rectangular silver box with the crest of the City of Leicester engraved in a central position on the lid. The three units are mounted on a rosewood stand located by a rectangular retaining wire of silver. The stand has an elevated base of rosewood which corresponds in shape to the centrally mounted box. Each of the glass inkwells is surmounted with a collar on to which a thick circular plate is attached by means of a flush hinge. The box is lined inside with rosewood and a section of the front of the box is cut away to expose the lining, thereby forming a thumb piece for opening the lid. Inside the box is a fitted blotter in rosewood with a flush adjusting silver screw, and resting in two channels in a platform, at a corresponding height to the blotter, are two pens, also in rosewood, with gilt nibs.

The glass was specially made by an optical lens manufacturer, being first prepared as cast blocks, and then machined with great accuracy to the finished dimensions. The hemispherical cavities for the inkwells were bored out and the finished glass highly polished. The precision and high finish of the glass makes an interesting contrast to the silver box, forming a counterchange of reflective transparent and solid forms.

The inkstand was given to the City of Leicester civic plate collection in 1960 as a gift from Lewis's Ltd, and was commissioned on behalf of Lewis's by George Tarratt of Leicester.

Made by John Limbrey 1960. Several of these ink-stands have been made; in 1964 for the Worshipful Company of Goldsmiths as a presentation gift to Lord Holford; in 1970 for Lucien Ercolani, Chairman of Ercol Furniture Ltd.

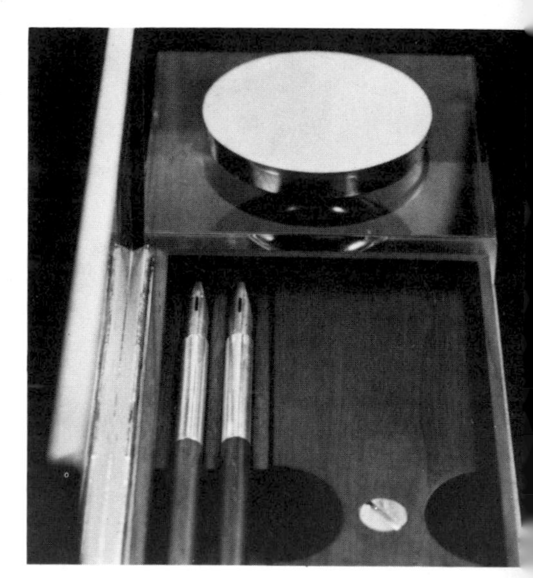

Coffee urn, cream and sugar bowl 1960

A one-gallon coffee urn mounted on a stand with a spirit heater, together with a matching sugar bowl and cream jug. The design was commissioned in 1959 by Sir John Cockroft, First Master of Churchill College, Cambridge.

The urn carries an engraved inscription and the crest of Churchill College engraved in a roundel above the tap. The swing handle, which is of silver, is inlaid with rosewood (see illustration) and has a special locking device so that the handle cannot pass beyond the horizontal position and thereby damage the silver body.

The body of the urn was made as a cone and hammered into the shape required. The stand was made from heavy-gauge wires; the top wire was rebated to take the stepped base of the urn, to avoid a pronounced demarcation between urn and stand. The base is made from laminated wood veneered with rosewood. The use of lamination overcomes the problem of warping caused by heat reflected from the base of the urn when the spirit heater is lit. The tap which is constructed entirely of silver, was made watertight by grinding the closure valve in position with grinding paste.

The complete set was made by John Limbrey in 1960.

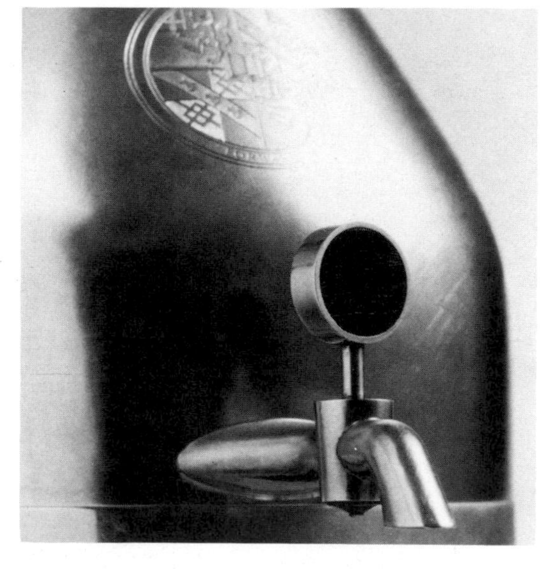

PRESENTED BY

This range of cutlery made in 18/8 steel includes more than forty different items. Illustrated opposite and overleaf are some of the pieces from the collection.

The knives have hollow handles with a concave section on both sides formed by pressing in two halves and argon welding them together. The blade, which is hot forged from a bar of stainless steel, has a pronounced bolster which is also argon welded on to the handle.

The fork and spoon handles are concave on the front and are made from blanks of slit stainless steel so that ends can be cross rolled to achieve a varying section along each piece. Both fork and spoon have a gently curved outline which follows through without a shoulder to the prongs or to the drop-shaped bowl.

The carving fork illustrated overleaf incorporates a spring-loaded guard made as a lost wax casting. This lies flat on the back of the fork when not in use. The cam action of the guard is operated by means of a stainless-steel spring set in a drilled hole in the back of the fork. To achieve the pronounced thickness required at this point, the fork is made as a hot forging from a bar of stainless steel.

Design work started early in 1961 and the place setting was submitted to Old Hall as hand-forged prototypes in stainless steel. Various modifications were made after user tests had been carried out, and a final hand-made set was prepared from which the dies were cut. Following the Campden pattern of 1957, this was the second range of cutlery to be made by Old Hall Tableware.

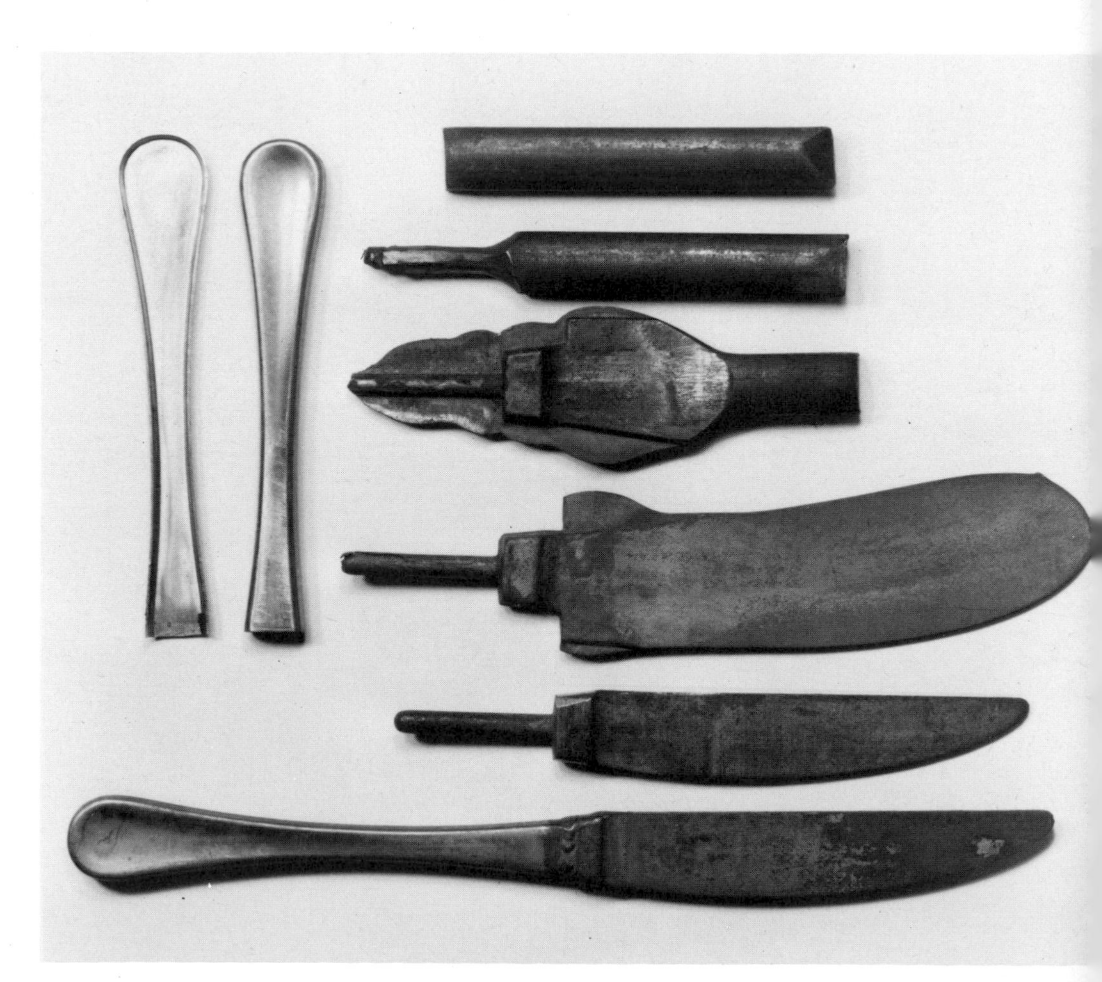

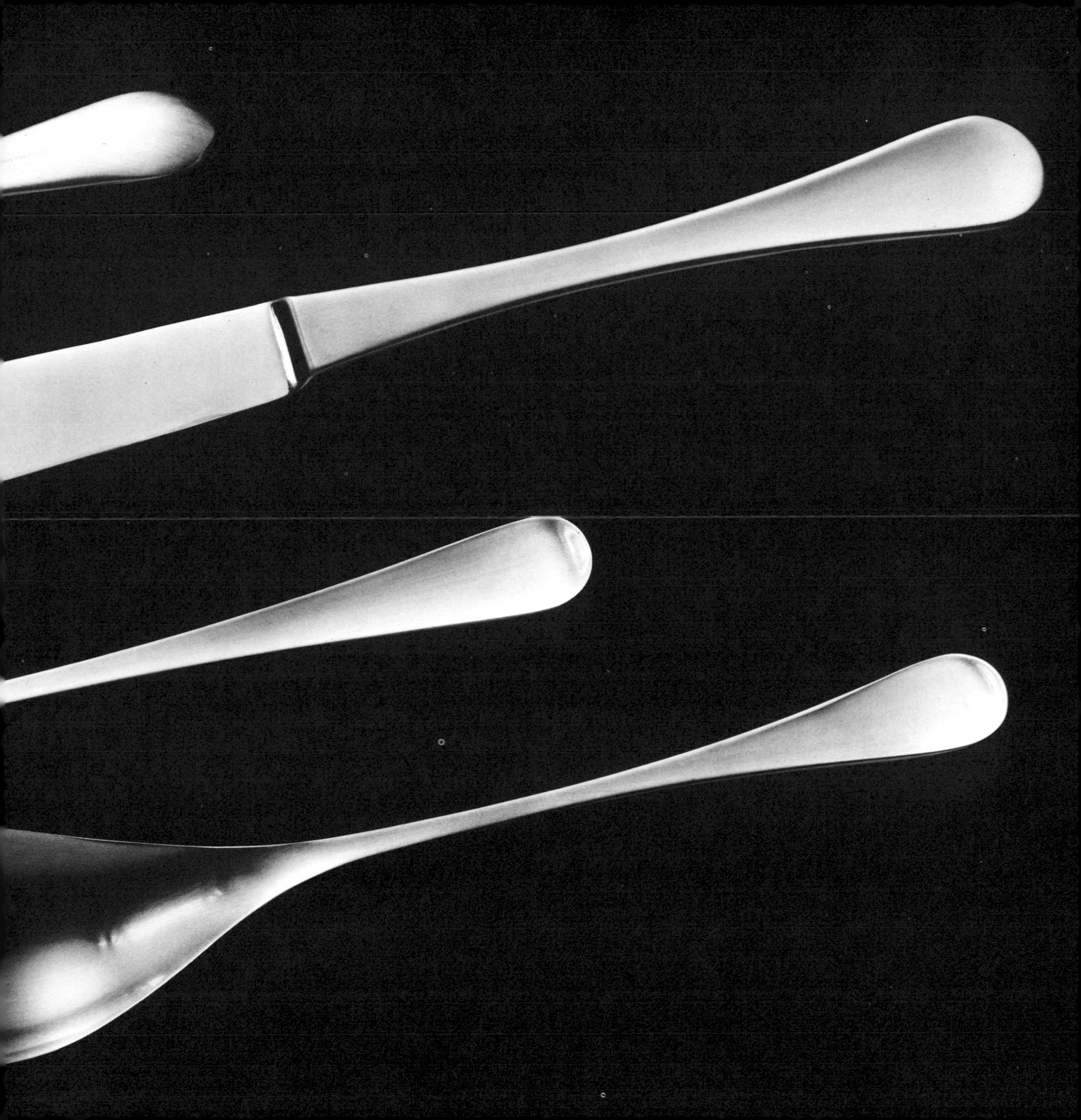

Knife, fork and spoon in stainless steel with rosewood handles. The stainless steel components are manufactured from the same thickness of metal, and all three pieces are made with a common tang fitting a deep concealed slot in the handle. The steel and rosewood are permanently secured by two compression rivets. A special high-speed miniature chain-saw cutter is used to make the concealed slot. This method has the advantage over other wooden handles, where the slot is exposed on the sides, that the handle can be completely finished off before assembly with a clear baked-on polyurethane protective coating.

The knife blade is mirror-polished, and is serrated to give a sharp cutting edge. The spoon is the only item to be cross-rolled before forming.

During the design stage several handmade models were produced in stainless steel and rosewood, and the cutlery was first made by Old Hall Tableware in 1963.

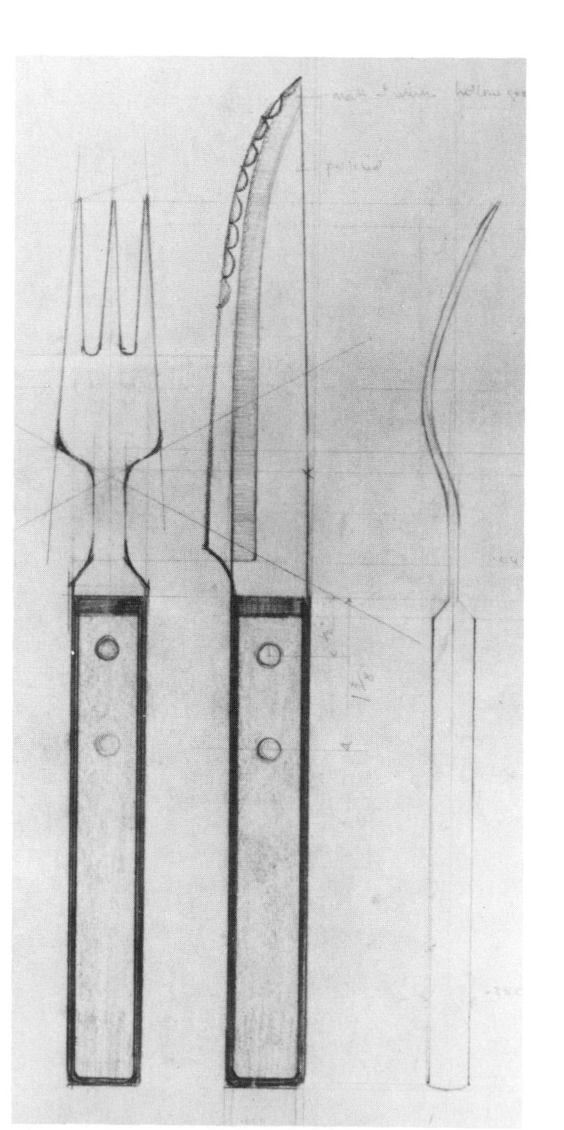

Cast iron, once used for many decorative products, has long been neglected. In 1962 the opportunity occurred to work in close collaboration with an iron foundry. The first design to be developed in this inexpensive material was a candlestick, to hold a 50mm diameter candle.

The distinctive feature of the design is the stem with its three wide discs. The casting was made by splitting the mould vertically and forming the candlesocket by using a core, which also made three pronounced ribs to take up variations in candle size. Various finishes were tried and eventually a black groundcoat, vitreous enamel fired at a temperature of 850°C, proved most satisfactory. This finish has a slightly granular quality which enhances the appearance of the iron.

Various designs were added to the candlestick to form a related range of products. These included a peppermill and a nutcracker in the form of a press, both designs reminiscent of early engineering forms.

At a later date the peppermill body was changed from iron to a cast aluminium in stove-enamel matt black, because of difficulties experienced in assembling the separate wooden base into the cast-iron body.

Marketed by Old Hall Tableware.

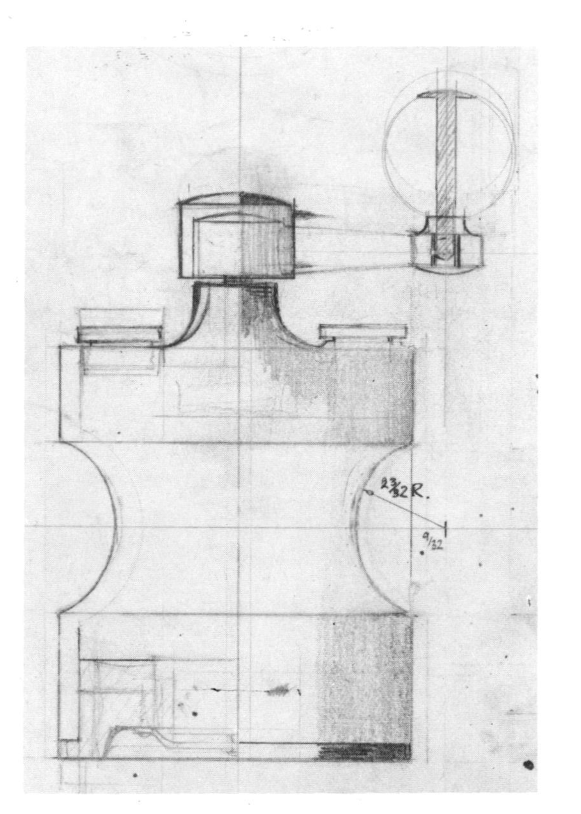

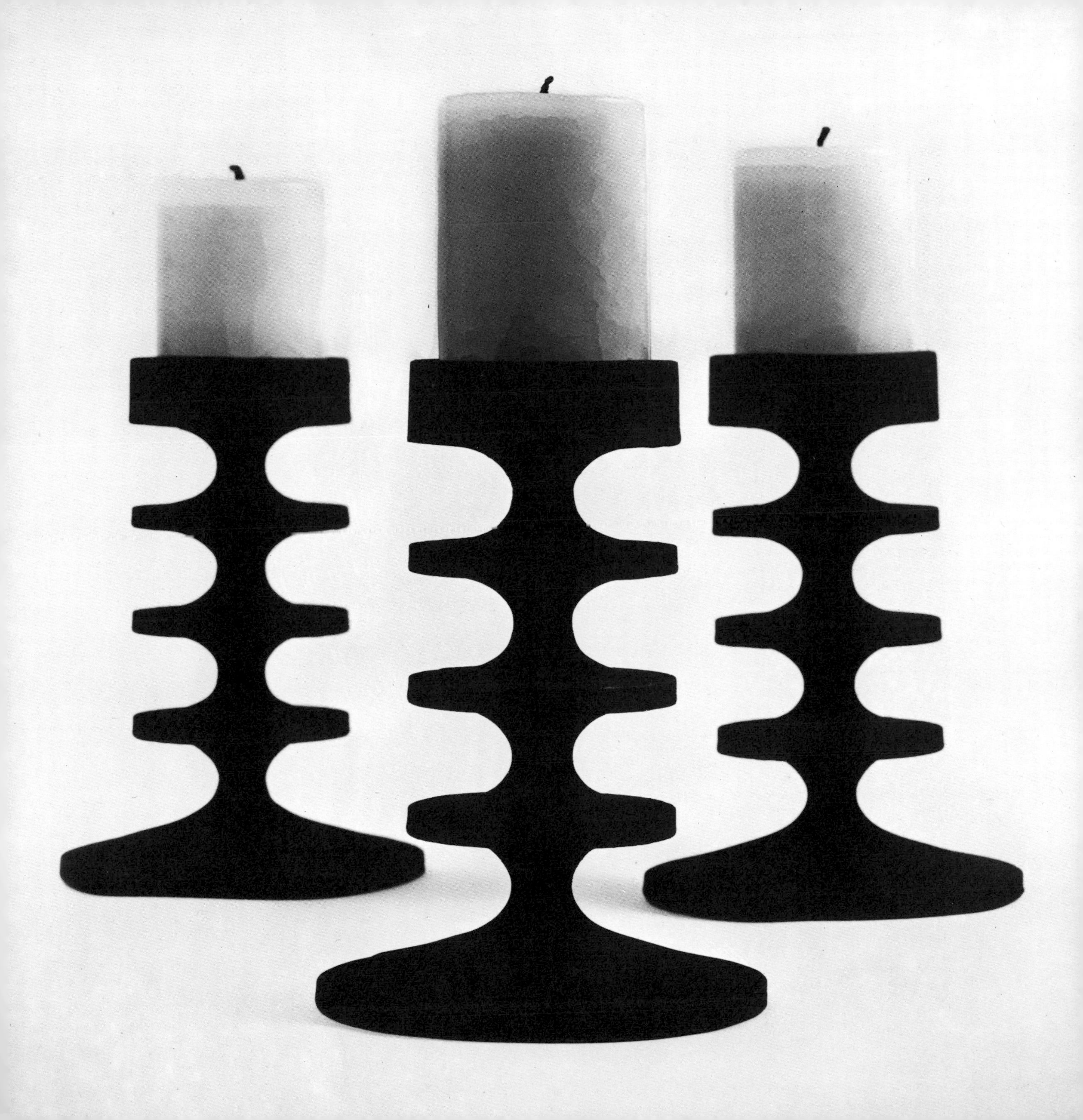

A stainless steel teapot of low, squat shape, which incorporated the use of lost wax castings for the spout and knob for the first time. This process was used by ancient civilizations: a pattern is made in wax and invested in a plaster or clay mould; when it is dry, a cavity is made by melting out the wax and this cavity is then filled with molten metal. In recent years it has been developed on a commercial basis for products in stainless steel. A group of spouts mounted on the runner system, before they are invested in the mould, is shown in the illustration.

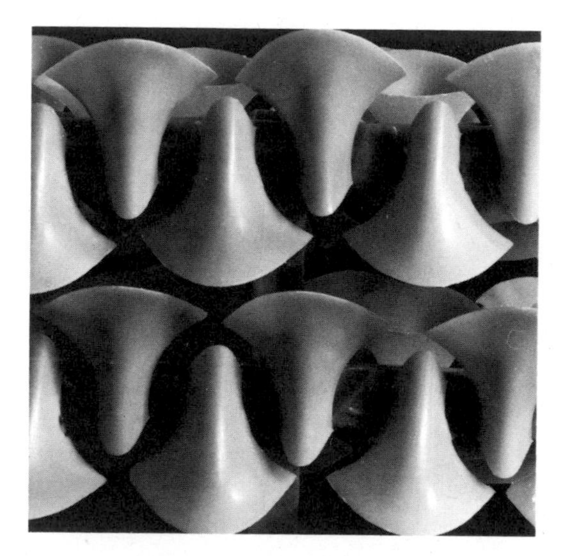

The use of a lost wax casting is particularly advantageous where great accuracy and good surface finish are required, and by using a casting for this spout it has been possible to blend a complex shape into the body of the pot. The spout is attached by argon welding which creates an invisible join after dressing and bobbing. The body is made from two pressings assembled by spot welding and infilled with argon welding, thus completely eliminating any trace of the join. The handle stays perfectly cool and is set at an angle which gives the user good balance and control when using the pot.

Manufactured by Old Hall Tableware.

In 1963 Heal's commissioned a range of table silver for sale in their 'Present Choice Department'. The collection was shown at a special exhibition at Heal's in August 1965, and since that time various additions have been made to the range.

This opportunity to make silver for sale through a retail outlet meant that the production costs were of great importance and therefore all the components were formed by spinning. This is a process of metal forming on a lathe to which is mounted a wooden or metal form of the desired shape. A flat circle of silver is burnished over the form by using long-handled blunt tools (see illustration). Sometimes this process leaves slight ripples on the surface of the silver and they are removed by gently hammering the surface, a process known as planishing.

The range was unified by the theme of a soft undulating curve reminiscent of the Churchill coffee urn, and rosewood was used for handles and knobs throughout.

On several occasions items from this collection have been chosen as Government gifts.

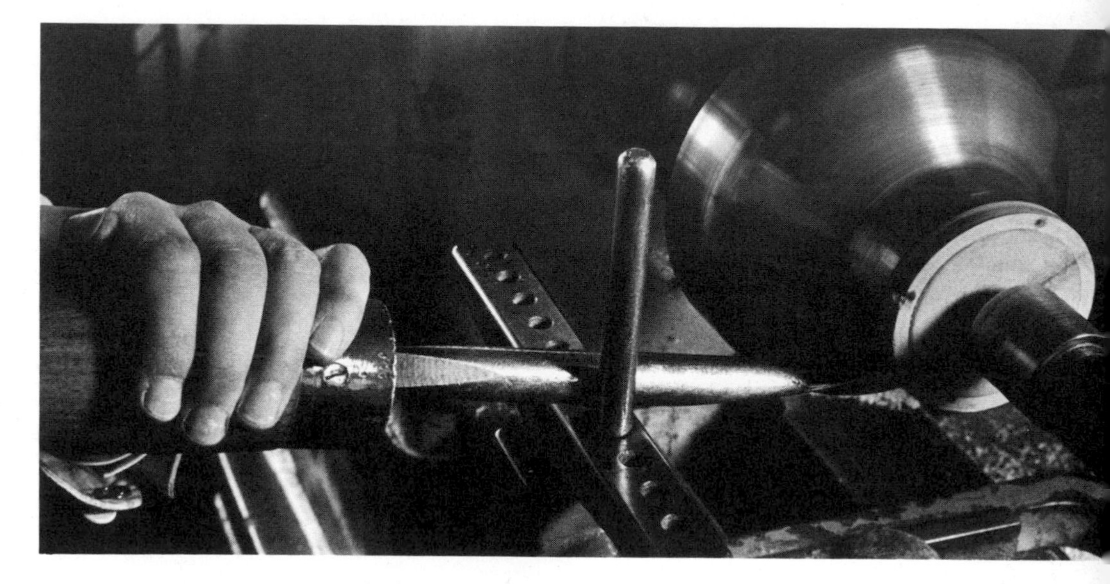

In 1929 Mr Chantry, a design and development engineer in Sheffield, saw a need for a device that would enable a knife blade to maintain a correct sharpening angle when applied to a steel; and after due experiment he invented and marketed a device that incorporated the best sharpening features of a steel with a completely foolproof system of operation. The device reproduces mechanically the movements used in sharpening knives with the ordinary butcher's steel: i.e. it brings the knife edge simultaneously along and across the steel; and ensures that the angle of the edge is correct.

Drawn through the machine, the blade moves against the two small steels mounted at this correct angle and under spring tension; this acts on both sides of the edge. The mechanism ensures that the optimum sharpening angle is maintained throughout the movement. The steels revolve in their holders and do not wear away the blade but roll it into a fine wavy line, thus giving a sharp cutting edge to knives, whether of stainless steel or ordinary steel, without doing any damage to the blades. The mechanism is still in use and unchanged to this day. In 1963 Harrison Fisher who had acquired the small company commissioned the design of a new case for this mechanism.

Previously the knife sharpener had two equal grips; the new design made had one large gripping area and the base was extended under the handle for stability. The case is die cast in mazak with a stove enamelled paint finish. The Chantry knife sharpener has been acquired for the Museum of Modern Art in New York and the Stedelijk Museum in Amsterdam.

Made by Harrison Fisher Ltd, Sheffield.

A pair of kitchen scissors made from high carbon steel chromium plated, overall length 220mm. The handles are finished in stove enamelled paint in a variety of colours. The scissors incorporate a device to loosen stubborn screw-down closures, and a bottle opener.

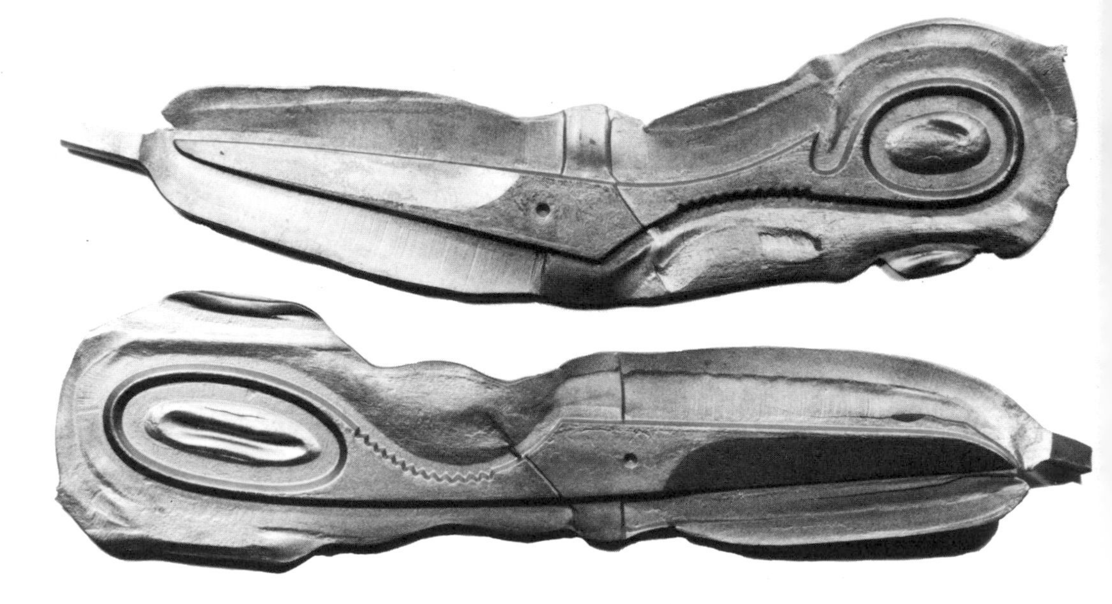

The design was first made as a study model in wood and then modified. Another model was made in brass with increased finger grips, and the general curvature of the lower line of the scissors was increased. Small modifications were made to the brass prototypes and working drawings were produced from which the dies were cut.

The first process of manufacture is hot forging the outline from material of dimensions calculated to fill the dies and provide an overspill or 'flash', the flash being removed by a press-work operation. Forging is done at a temperature of approx. 800°C for carbon steel scissors.

The finished blank is hardened and tempered under accurately controlled conditions to provide prolonged working life. The blades are then ground, glazed and finished. It is important that the inside of the blade should be hollow ground because a correctly functioning scissor is designed so that top and bottom blades are in contact with each other only at a single point which moves up and down the length of the blades as they are opened and closed. After grinding and glazing the blades are set in a skilled hand operation known as 'putting together'. Well-made scissors cut smoothly from 'pinch to point' i.e. from beginning to end of the blades; they work smoothly and cut without the user having to press one blade against another. This is called 'holding on the cut'.

Made by Taylors Eye Witness Ltd, Sheffield.

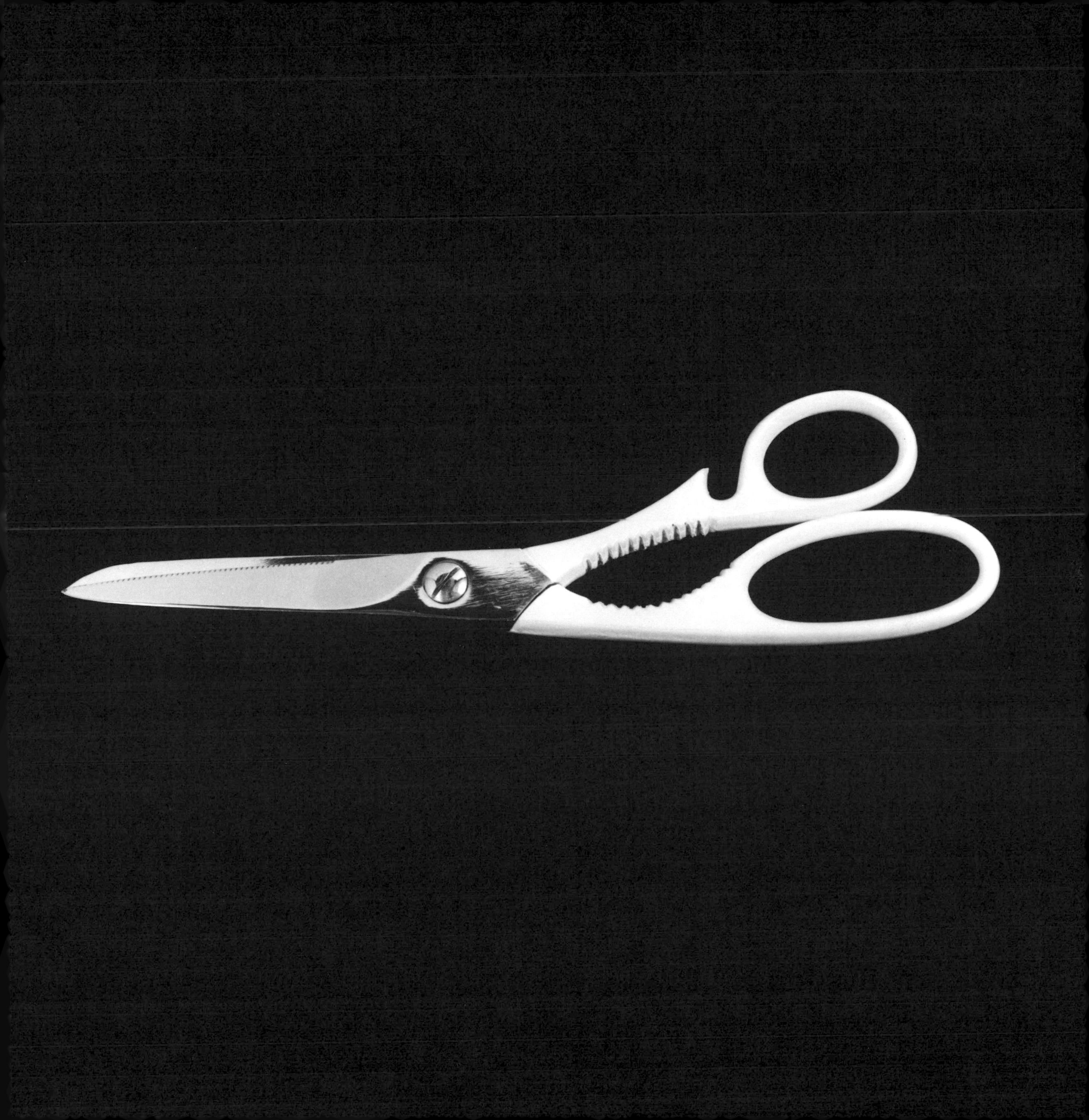

A table lamp of 315mm diameter, in acrylic and polished aluminium. The lamp is formed from two domes of opal acrylic, the upper surmounted by an outer dome of smoke-coloured acrylic which effectively diffuses the light. The acrylic assembly is separated above and supported below the lamp by two identical aluminium spinnings pierced with a circle of ventilation holes. The lower spinning houses the bulb holder and the switch, which is operated with a ball chain. The base of the lamp, which is inset with a cast-iron weight, is formed by a spinning joined to the vertical tube. These components require very few modifications to make a pendant lamp version, and the principle of interchangeable components is used throughout. For instance, the outer domes for the smaller lamps use the same tooling as the inner domes for the larger ones.

The range consists of two sizes of table lamp, a floor standard, two sizes of pendant, and a wall bracket. The lamps were first shown at Lumitron in the form of a working model of the 315mm table lamp made in brass, and were put into production after only minor changes had been made.

Made by Lumitron Ltd, London.

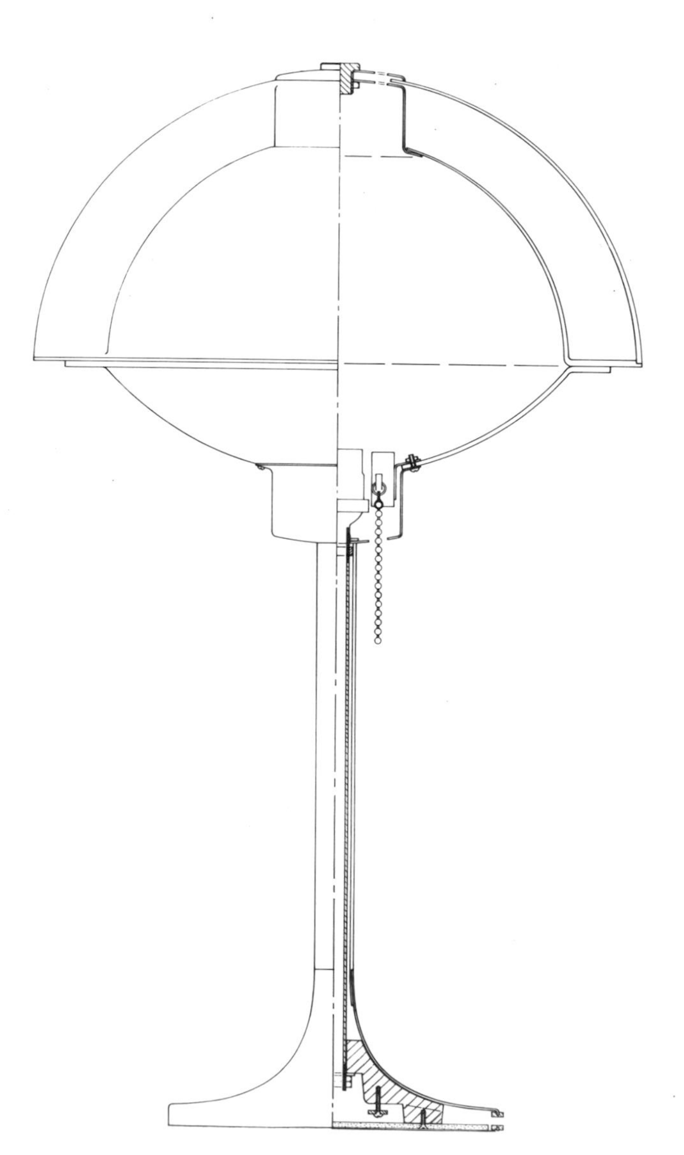

Kettle in enamel steel 1966

Kettle in enamel steel of $2\frac{1}{2}$ litres capacity. The lid is of stainless steel and the handle and knob of black plastic. The knob is mounted offset from the centre, opposite a slot in the vertical face of the lid which acts as a steam vent. The low, squat, shape is particularly suitable for use on gas and electric cookers.

The long curved handle, a soft 'D' shape in section, relates to the shape of the body. This is formed from two pressings welded together by automatic argon arc welding. The design was shown as a study model in wood and a working model in gilding metal. Only minor changes were made before it was put into production. The kettle is finished in red, black or dark blue vitreous enamel with black enamel inside.

The kettle was the first commission to be undertaken for this West German firm; subsequently a range of related saucepans, casseroles and a coffee pot were added to the collection. Exhibited at the Stedelijk Museum, Amsterdam, and acquired for their permanent collection, 1968.

Made by Carl Prinz AG, Solingen, West Germany.

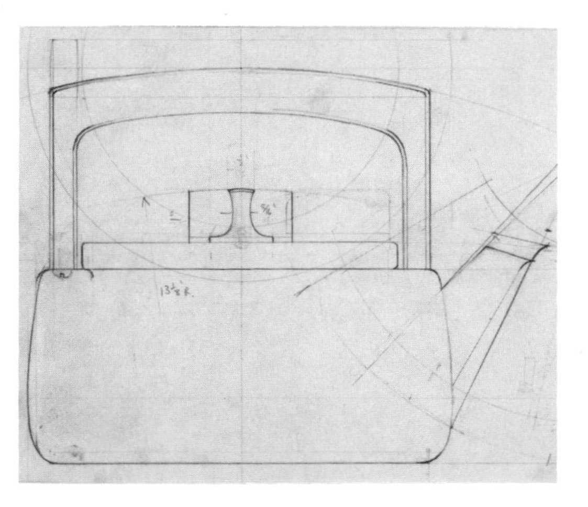

Adjustable lamp made in three sizes (two sizes of table lamps and a floor standard). The base of the lamp is made from steel with a stove enamel finish and a cast-iron weight inset for stability. The heavy-duty, chrome-plated, flexible hose is attached to the stem of the base with a collar which enables the hose to rotate through 180 degrees against stops, which prevent the wires from becoming twisted. The shade is spun from aluminium, mirror finished and anodized. A baffle and louvre are fitted to the shade to prevent glare.

The design was the outcome of research into the use of industrial hose, and the table lamp was shown to Lumitron as a working model developed from the original drawing (see illustration). Some modifications were necessary before the lamps were put into production; the original cast-iron base was changed to sheet steel, and the length of the hose reduced slightly.

Made by Lumitron Ltd.

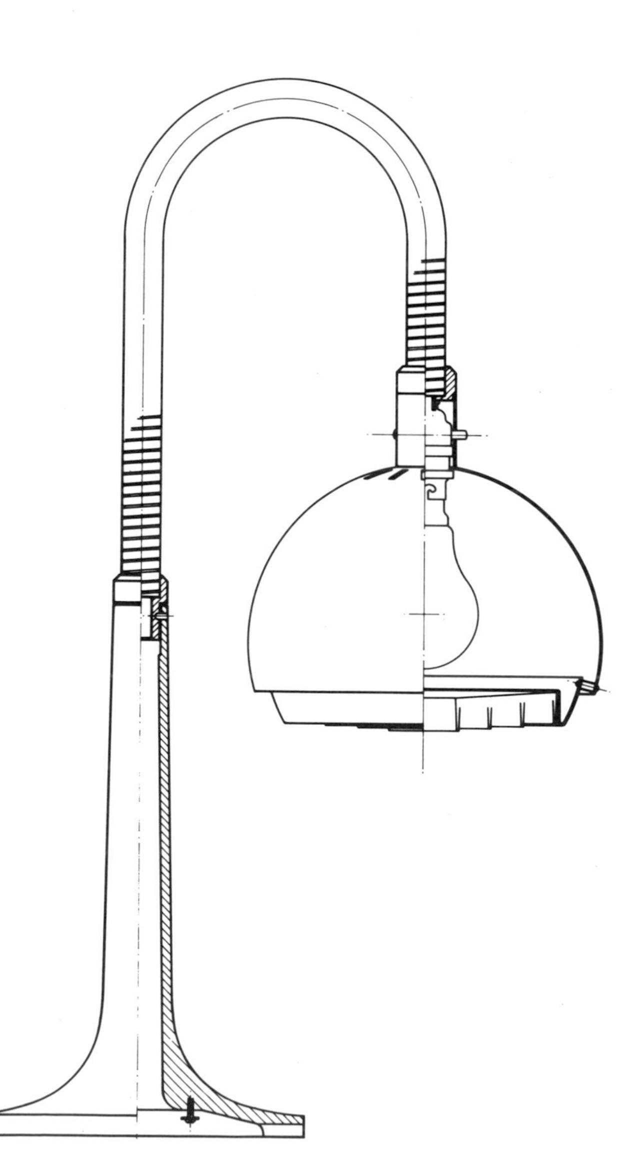

This candelabrum was constructed by attaching six vertical rods to the raised central boss of the base. An internal chassis system secures them in accurate alignment. In the centre of the boss the crest of the Society is engraved in the form of a roundel, and an inscription referring to the gift of the silver is engraved round the rim of the base.

The rods curve outwards in a progressively flattening section, and the candle sockets and sconces are mounted near the ends on short lengths of rod. When the design was prepared it was hoped that the rods could be forged to form the curved tapering flat required at the extremities. However, it proved impossible to achieve the desired shape in this way, so the curved section was made as a casting in silver. It was then soldered on to the rod and filed smooth.

This candelabrum, one of a set of six, was commissioned by the Royal Society of Arts together with matching condiment sets in 1965. A special feature of the design is that the candles are set high above the level of the table top to avoid restriction of vision for the seated diners.

Made by Raymond Marsh. The weight is 102oz, height 460mm, width 290mm.

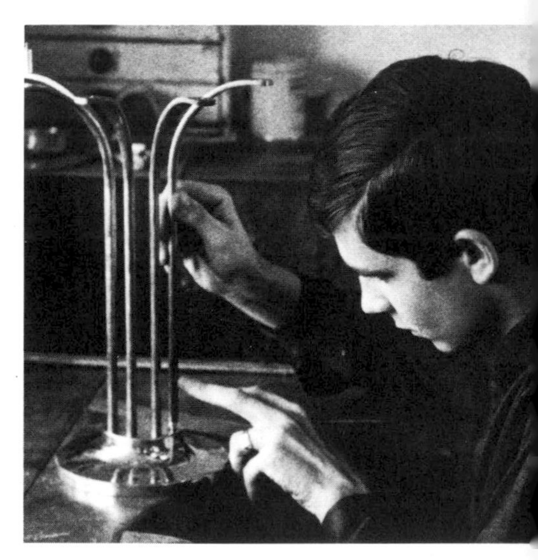

The goblet illustrated is made of full lead crystal glass, a heavy material which naturally forms itself into soft, rounded shapes in working. They are entirely hand blown, and the design expresses some of the essential characteristics of lead glass.

Before design work began several museums were visited, and studies were made of late eighteenth-century English tavern glasses. These, with their robust appearance, suggested the development of a new design. Considerable time was spent in the glass house watching the glass blowers at work, and finally a set of models was made in solid acrylic. These were used as patterns by the glass blowers.

The decanter in lead crystal is formed from a wide flattened sphere which gives it the characteristic 'Ship's decanter' appearance. This is echoed in the silver mount, formed from two wide flanges, the top flange being the silver stopper. The lower flange is formed from an internal collar which spreads outwards in a shallow curve. The glass is prepared for the silver mount by cutting deep grooves in the concealed area, and plaster of paris is used to cement the two parts together.

Goblets and decanter made for Robert Welch by the Bridge Crystal Glass Co., Stourbridge (a member of the Prestige Group).

RW

The two large silver candelabra were specially commissioned by the Worshipful Company of Goldsmiths for use in Goldsmiths' Hall. The magnificent hall and splendid setting called for a design of great richness, especially as the candelabra would stand on certain occasions next to silver by Paul Storr, the famous nineteenth-century silversmith.

The sketch submitted to the Wardens showed a proposal for a candelabrum covered with small hemispherical forms designed to have a shimmering effect by candlelight. The sketch was accompanied by a section of the main stem turned from brass rods and fitted around a supporting chassis.

The design involved an intricate method of construction and it would have been totally impractical to have made the interlocking spherical forms from solid silver because of the weight factor.

Various experiments were tried with swaging tube, a method of shaping tube between reciprocating dies which hammer the tube into the desired shape. These were abandoned because of the limited depth of the swaging available on a conventional machine. Eventually, by good fortune, a machine was located that had been made at the turn of the century to form the balls for hunter watch winders. It was found that with a pair of special dies, formed tube to the required length and interlocking pitch could be produced. The candelabra were constructed with an elaborate chassis system so that the tubes could be supported and fitted into position by a special locking device. Each of the tubes was highly polished and hard gold plated before being assembled.

The two candelabra were made in 1970 by John Limbrey and Paul Heneghan; the finished weight of each 416oz, height 762mm, width 559mm.

The illustrations show three items of silverware using a repoussé technique to form a 'cobblestone' surface texture. This approach is an interpretation of the three-dimensional decorative effect achieved on the large candelabra made for the Worshipful Company of Goldsmiths.

The first piece made using this technique was a large shallow circular dish 425mm diameter, commissioned by Dr Armstrong as a gift from Churchill College, Cambridge. When the design was submitted various techniques, such as casting and electro-forming, were being considered. It was eventually decided, however, to inlay into the rim a thin sheet of silver having the smooth rounded domes punched up from the back. The crispness of the texture was produced by using a heavy brass former drilled with a specially rounded drill to make a negative pattern of the desired effect. Having spun the thin sheet over the former, to the rim profile, a steel punch was used to drive the silver home into the holes.

A similar technique was used to make the 83mm square box. The developed shape required to form the box from a single sheet of silver was drawn on a flat brass plate and the pattern drilled into the surface. A sheet of silver was fixed to the brass die with adhesive tape, and a hide mallet used to start the repoussé, giving witness marks on the surface which could be followed with a steel punch. On soldering the scored and folded corners of the box it was found that the texturing of the sheet greatly reduced the problem of distortion often encountered in box making. Another advantage of this decorative effect is that articles which are handled in use, as in this little box, do not show finger marks and need cleaning far less frequently.

The nearly hemispherical rose bowl illustrates another variation of this technique. As the 'cobblestone' pattern, in the form of clusters of domes of varying size, is placed on the near-vertical sides of the bowl, the negative die had to be made as a four-piece mould in cast iron. The silver bowl was shaped to be a close fit in the mould and the pattern punched into the drilled holes as before. Used in this way the reverse side of the pattern is now clearly visible, as a decorative effect.

The cover is a slightly domed plate pierced with irregularly spaced circular holes of varying size. Each of these holes is surrounded by rings of repoussé domes.

Coffee pot, condiment set and oil bottle, from a range of stoneware made by Brixham Pottery in their seventeenth-century Pound House, near the famous Devon harbour.

The coffee pot is 240mm high and has a capacity of one litre. The body of the pot is cylindrical, and the sides slope inwards from the wide base to the taller, narrower, neck. The lid is a hemispherical dome with a pronounced rounded lip. A special feature of the design is the compactness of the handle and spout, both of which are set into the narrow part of the body, thus reducing the over-all projection of the pot.

The oil bottle, 160mm high, is related in shape to the coffee pot, but has no handle; the condiment set is a simple cylinder with a domed and lipped top.

The stoneware is finished in Tenmoku glaze, which gives a warm mottled chestnut colour, with tinges of red. This glaze does not lend itself to uniform repetition, and considerable variations in colour are produced.

Design work started in 1970 with a coffee service. Drawings and study models were made and Brixham Pottery produced the blocks, cases and moulds necessary to make the range.

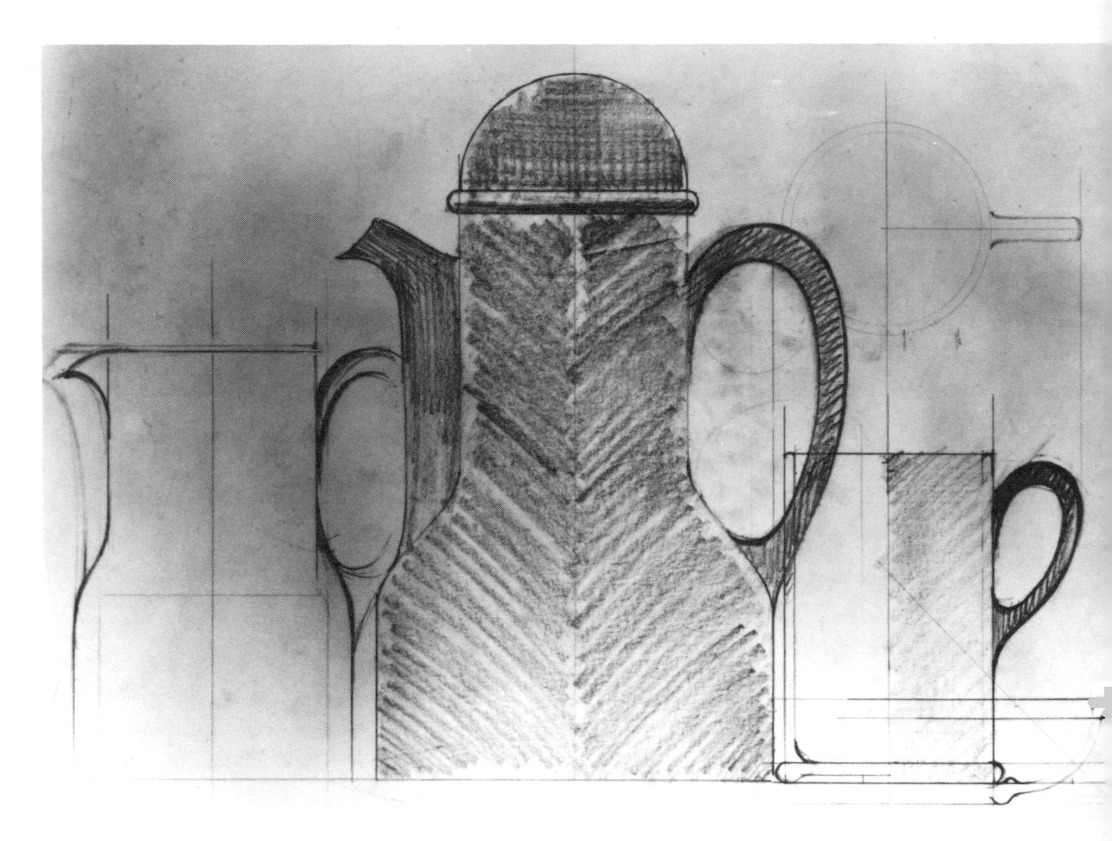

This range of cast-iron cookware was commissioned in 1970 by H. E. Lauffer Inc., New Jersey, USA. Before design work started, the requirements of the American market were studied and a careful brief was prepared with a special emphasis on function. For instance, the lids of all the casseroles had to serve as both cooking and serving pieces.

Initially, the range was planned as a set of round and oval casseroles with frypans, the designs being so related that the range could be considerably expanded as new designs were developed. A special feature of the design is the two ribs around the casserole from which the handles were developed; the lower rib is set at the point where the pattern for the base of the casserole divides into two parts for moulding. Any extra metal, in the form of flash created during the casting process, can readily be ground off from the rib. The open handles are extremely functional in use, and are also important from the production point of view, as the lids and bases can be hung in the vitreous enamelling process. The finish of all pieces is vitreous enamel, semi-matt black outside and fully acid resistant self-mottled gloss grey inside.

Made in England by Qualcast Ltd, Wolverhampton, and sold world-wide by companies associated with Lauffer.

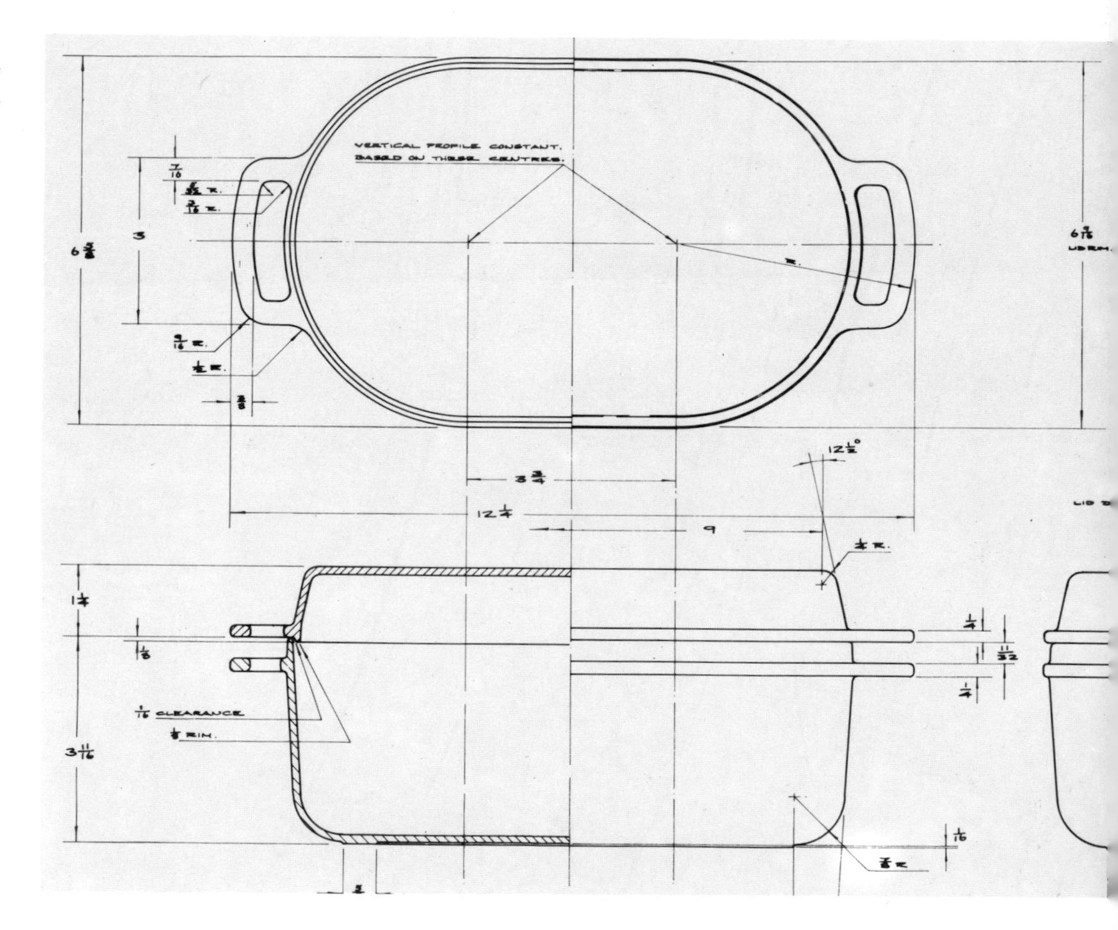

The body of the pot is constructed by forming a seamed oval cone and soldering this to a hand-raised base, the assembled pot is then planished to perfect the shape (see illustration).

The lid is constructed from a flat plate inlaid with a flush hinge. This type of hinge is achieved by using thick walled knuckles and filing away the top protruding parts. The resulting smooth, flat finish to the lid contrasts with the hammered finish of the body. The handle is sawn in profile from a section of African elephant tusk and is shaped with a file until it fits the silver sockets on the body. The knob is also filed into shape and fitted into a socket riveted to the lid.

A larger size of teapot, matching coffee and hot water pots, cream jugs and sugar basins, have been made.

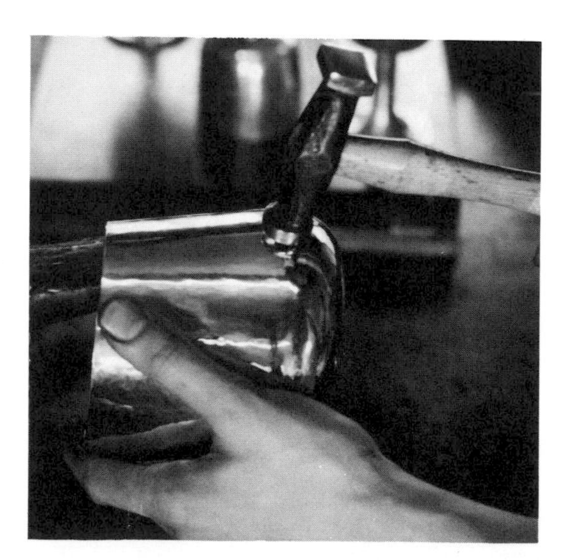

A large circular bowl, 500mm diameter, slightly domed at the centre, with a planished finish all over. The upper portion of the inside is parcel gilt. The gilt area is decorated with applied wires running from the rim towards the centre of the bowl in rhythmic undulating forms. Amethysts are set at the ends of many of the wires, while the rest terminate in small beads.

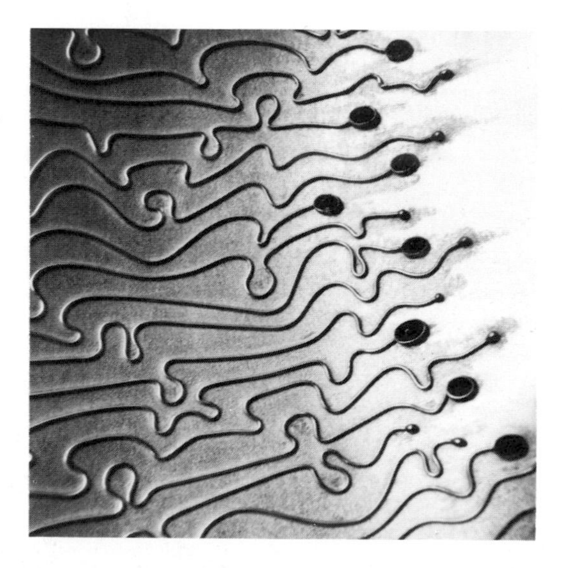

The bowl was made in silver of 1mm thickness with a rounded, rectangular section wire soldered on to the edge. The area to be decorated with applied wires was calculated and drawn on paper as a flat segment. Rough outlines for the wires were drawn within the segment as a guide before the wires were formed. Each wire was bent with pliers and held in position with adhesive tape as the varying rhythmic patterns were developed. When the segment had been completed the position of the amethysts was finalized, all wires that terminated without stones had the ends melted to form small beads. The wires were then formed to match the curve of the bowl and soldered in position, in small batches, on the textured area. Settings were then made for the amethysts and these were soldered in position, the stones being set after the gilding process.

Made by John Limbrey 1973.